WISCONSIN'S CHAMPION TREES

A Tree Hunter's Guide

by
R. Bruce Allison

Photographs by
B. Wolfgang Hoffmann

Foreword by
Dick Rideout

Wisconsin Book Publishing  Verona, Wisconsin

ACKNOWLEDGEMENTS

Printing of Wisconsin's Champion Trees was made possible through generous contributions from the following:

Dane County Tree Board
Madison Gas & Electric Foundation, Inc.
Alliant Energy Corporation
We Energies Corporation Foundation
Hooper Foundation
Terry Monson

⚓

Front cover photo: Eastern poplar located in Sharon Township, Walworth County, B-*Wolfgang Hoffmann (2004*

Published by Wisconsin Book Publishing
1830 Sugar River Road, Verona, WI 53593
wisconsinbookpublishing.com
ISBN 0-913370-18-5

TABLE OF CONTENTS

Dick Rideout
Official keeper of the Wisconsin champion tree registry

FOREWORD

MY INTEREST IN BIG TREES probably started with the giant American elm that shaded our entire back yard in Evanston, IL where I was born. When I was five, my family spent a year in California and we visited the redwoods and sequoias. It's hard not to become a big tree lover after you've seen these incredible creatures, even if you're only five.

Without knowing it, these early experiences probably drew me to the field of horticulture at the University of Wisconsin and my major professor, Dr. Ed Hasselkus. I vividly recall measuring trees with Ed in the mid 70's and finding my first state champion with him, a fastigiate ginkgo in Milwaukee. In the 80's I worked for the city of Milwaukee Bureau of Forestry and saw an opportunity to excite the staff with a champion tree contest. When it was over, we had found 18 new state champions in Milwaukee alone!

In 1990, I moved back to Madison to develop Wisconsin's state urban forestry program at the Department of Natural Resources. At the time, the DNR maintained the official records of Wisconsin's native trees and Ed Hasselkus maintained those of the non-native trees. Upon my arrival at DNR, a box of records mysteriously appeared on my desk and I became the official keeper of the native champion trees. And then in 1994, Ed Hasselkus retired and with his vacant position in possible jeopardy, the non-native records arrived on my desk as well. The DNR was now the repository for both native and non-native champion tree records.

As the official registry of all of Wisconsin's largest trees, the DNR receives nominations from interested citizens from across the state. Volunteer "big tree inspectors" must verify the measurements and species before the information is entered in the official record book. The DNR maintains information on the location, dimensions and rank of Wisconsin's largest specimens of over 270 tree species and cultivars, totaling over 2200 records. This registry is published occasionally, but is always available on the DNR's web site at: *http://dnr.wi.gov/org/land/forestry/uf/champion/*. Information on how to measure and nominate new champions is there as well.

The state champion tree registry is the result of the work of many people and reflects the great interest Wisconsin citizens have in our largest trees. We depend upon this citizen interest to keep the records current with new information and measurements. We are very grateful to Bruce and Wolfgang for the effort they have put into updating and photographing many of the largest trees for this book.

In keeping these records, DNR hopes to promote the appreciation of Wisconsin's forests and trees. I encourage you to participate by hiking into our state woods and urban forests to enjoy the majesty of our current champion trees and to seek out and submit nominations for new ones. And take your children with you. They just might be inspired like I was.

Dick Rideout
State Urban Forestry Coordinator
Division of Forestry
Wisconsin Department of Natural Resources

PREFACE

AS A PROFESSIONAL CONSULTING ARBORIST I make my living diagnosing trees and advising on how to keep them healthy. I love my work in part because I never cease to be amazed at the remarkable nature of trees. Trees are the oldest, tallest, most massive organisms on earth. My patients can be hundreds of years old, reach over 100 feet skyward and weigh many tons.

Whenever I come across a tree that is unusually large for its species, I will pull out my tape measure, wrap it around the trunk and compare its circumference to those recorded in the state record of big trees. My friend and mentor, the late Walter E. Scott, was the first caretaker of the Wisconsin record of big trees. As assistant to the Secretary of the Department of Conservation, he initiated the program in 1941 following the lead of the American Forestry Association national record of big trees. Another important teacher in my life, University of Wisconsin Horticulture Professor Emeritus Edward Hasselkus, maintained records of non-native champion trees from 1977-1994. It has been 25 years since Walter and Ed encouraged me to publish a general audience version of those records titled *Wisconsin's Champion Trees* (Wisconsin Books, 1980). That book was well received. People tell me they still carry dog-eared copies in the glove box at the ready to make comparison to any new tree discoveries.

Recognizing it was time for a revision, I selected the champion and top contenders from the state records now maintained by the Wisconsin Department of Natural Resources (DNR) Division of Forestry. With the help of my staff Sarah Weishaupt, Joanna Buckner and Travis Simmer, hundreds of phone calls and visits were made in an attempt to correct and update selected records. We could not have succeeded without the cooperation of tree owners, DNR, University and Extension personnel plus many enthusiastic fellow big tree hunters throughout the state who joined in the effort.

Trees are dynamic, growing and dying, within an ever-changing environment. New trees have replaced many fallen champions. Although we worked hard for over a year to revise old records, many of the trees listed are still in need of updated information. I pass the challenge onto the reader to locate and measure them.

In this book I rank the trees by the most reliable and easily measured statistic, the trunk circumference measured at breast height (**CBH**). The official state ranking differs occasionally because it uses a point system with height (**H**) and crown spread (**CS**) in addition to circumference. I list all three measurements when they are available. Also listed, when available, are the dates and persons responsible for the nomination and most recent measurement. The best location information available is provided. Tree species are listed alphabetically by scientific name. A glossary in the front of the book allows the reader to easily translate common names into scientific names. Plus, a common name index in the back facilitates locating tree species by page number. The Division of Forestry recommendations for measuring and nominating trees are also included.

I was fortunate to have the photographer from the first edition, my close friend B-Wolfgang Hoffmann, photojournalist at the UW-Madison College of Agricultural and Life Sciences, join me in this new adventure.

Many of the trees in the state records are still to be remeasured and of course there are many champion trees "hiding" in the woods, backyards or along streets waiting to be discovered, measured and recorded. It is my fervent hope that my efforts in publishing this book will re-invigorate seasoned big tree hunters to return to the search. I hope it will also stimulate a new generation of champion tree hunters to enter the field, place their hands on the trunk, look up and realize what a valuable resource we have in our state trees. I hope you find as much enjoyment in using this book as I had in putting it together.

R. Bruce Allison
January 1, 2005

P. S. A humorous quotation on the tradition of big tree hunting by famous American author, Oliver Wendell Holmes, conversing with fellow boarders from his 1858 book *The Autocrat of the Breakfast Table*:

"I wonder how my great trees are coming on this summer?"

"Where are your great trees, Sir?" said the divinity-student.

"Oh, all around about New England. I call all trees mine that I have put my wedding-ring on, and I have as many tree-wives as Brigham Young has human ones."

"One set's as green as the other," exclaimed a boarder, who has never been identified.

"They're all Bloomers," said the young fellow called John.

(I should have rebuked this trifling with language, if our landlady's daughter had not asked me just then what I meant by putting my wedding-ring on a tree.)

"Why, measuring it with my thirty-foot tape, my dear," said I, "I have worn a tape almost out on the rough barks of our old New England elms and other big trees."

HOW TO MEASURE A CHAMPION TREE

CIRCUMFERENCE MEASUREMENT

Using a flexible tape measure, measure the distance around the trunk of the tree to the nearest inch. This measurement should be taken at 4½ feet above ground level. If the tree is on a slope use the midpoint of the tree base to measure 4½ feet above ground level. If there is a branch or growth on the trunk at 4½ feet measure the circumference just below the obstruction and report the height at which the measurement was taken. For multi-trunked trees that branch below 4½ feet report the circumference of the largest trunk at 4½ feet. If a multi-trunked tree flares out at 4½ feet measure the smallest circumference below 4½ feet and report height at which the measurement was taken.

HEIGHT MEASUREMENT

Take a 12-inch ruler and hold it vertically at eye level, in an outstretched arm. Stand far enough away from the tree so that you can roughly see both the base and the top of the tree between the top and bottom of the ruler. Move forward or backward until the eye sights the base of the tree across the 0-inch gradation and the tip of the crown across the 10-inch gradation. Then a sight is taken across the one-inch gradation and a companion marks the corresponding point on the tree. Using a tape measure, measure the distance from the base of the tree to this point to the nearest foot and multiply by ten. This is the height of the tree. If a height measuring instrument is available, its use is preferred. Be sure to report your method of measurement and have someone else verify your results.

CROWN SPREAD MEASUREMENT

Place a marker under the outside edge of the crown that is farthest from the trunk and another directly opposite at the crown. Next, set a marker at the edge of the crown that is closest to the trunk and another at the outer edge of the crown directly opposite it. Using a tape measure, measure both distances to the nearest foot. Add the two measurements together and divide the sum by two to obtain the average crown spread.

POINT VALUE

The total point value, according to American Forests, is calculated by adding circumference in inches, the height in feet and ¼ of the average crown spread in feet.

HOW TO NOMINATE A CHAMPION TREE

Complete a **DNR Champion Tree Program Nomination Form** *(available upon request from the DNR or online at www.dnr.state.wi.us/org/land/forestry/uf/champion/ nomination.pdf)* to record the following information:

Species – Species identification, both common and scientific name, including genus, species, and variety or cultivar designation. Refer to a good tree identification book and, if necessary, take a twig and leaf specimen to an authority for positive identification.

Tree Measurement – Circumference in inches, height in feet and average crown spread in feet. Note the height at which the circumference was measured if different from 4¹/₂ feet. A Big Tree Inspector or a professional forester must verify measurements.

Exact Location of the Tree – Include county, and municipality or township. In rural areas, report the section number, quarter section, and location in relationship to highways, farmsteads, woodlots or other geographic features. The use of global positioning system (GPS) co-ordinates is preferred. In urban areas report the street address and an indication of the location of the tree on the property, such as backyard or street tree.

Other Information – Names, addresses and phone numbers of the measurer, owner(s) and nominator, date of measurement, comments on the tree's condition such as: age, history, health, and/or other relevant information.

SUBMITTING A NOMINATION
A Big Tree Inspector or professional forester must verify all measurements. Submit the nomination to one of the inspectors who covers the area where the tree is located. Once the inspector has checked the measurements and signed the nomination form, submit the form to:
Champion Tree Program, Department of Natural Resources
P.O. Box 7921
Madison, WI 53707

SUBMITTING UPDATES
If you are remeasuring a tree, submit a letter to the above address listing the exact identity of the tree, the new verified measurements, and any other new information about the tree's age, condition, location, owners, etc.

GLOSSARY OF COMMON AND SCIENTIFIC TREE NAMES

Common Name	Genus Name	Common Name	Genus Name
Alder	*Alnus*	Katsuratree	*Cercidiphyllum*
Apple	*Malus*	Kentrucky Coffeetree	*Gymnocladus*
Apricot	*Prunus*	Larch	*Larix*
Arborvitae	*Thuja*	Lilac	*Syringa*
Ash	*Fraxinus*	Linden	*Tilia*
Aspen	*Populus*	Locust	*Robinia*
Baldcypress	*Taxodium*	Maackia	*Maackia*
Basswood	*Tilia*	Magnolia	*Magnolia*
Beech	*Fagus*	Maple	*Acer*
Birch	*Betula*	Mountainash	*Sorbus*
Boxelder	*Acer*	Mulberry	*Morus*
Buckeye	*Aesculus*	Oak	*Quercus*
Buckthorn	*Rhamnus*	Oriental Arborvitae	*Platycladus*
Butternut	*Juglans*	Osageorange	*Maclura*
Castor-aralia	*Kalopanax*	Pawpaw	*Asimina*
Catalpa	*Catalpa*	Pear	*Pyrus*
Cedar	*Juniperus*	Pecan	*Carya*
Cherry	*Prunus*	Persimmon	*Diospyros*
Chestnut	*Castanea*	Pine	*Pinus*
Chokecherry	*Prunus*	Planetree	*Platanus*
Corktree	*Phellodendron*	Plum	*Prunus*
Cottonwood	*Populus*	Poplar	*Populus*
Crabapple	*Malus*	Prickly-ash	*Zanthoxylum*
Cucumbertree	*Magnolia*	Quince	*Cydonia*
Dawnredwood	*Metasequoia*	Redbud	*Cercis*
Devil's Walking Stick	*Aralia*	Russianolive	*Elaeagnus*
Dogwood	*Cornus*	Sassafras	*Sassafras*
Douglasfir	*Pseudotsuga*	Serviceberry	*Amelanchier*
Elm	*Ulmus*	Silverbell	*Halesia*
Euonymus	*Euonymus*	Smoketree	*Cotinus*
Falsecypress	*Chamaecyparis*	Spindletree	*Euonymus*
Filbert	*Corylus*	Spruce	*Picea*
Fir	*Abies*	Sumac	*Rhus*
Fringetree	*Chionanthus*	Sweetgum	*Liquidambar*
Ginkgo	*Ginkgo*	Sycamore	*Platanus*
Goldenraintree	*Koelreuteria*	Tamarack	*Larix*
Hackberry	*Celtis*	Tamarix	*Tamarix*
Hawthorn	*Crataegus*	Tree of Heaven	*Ailanthus*
Heartnut	*Juglans*	Tuliptree	*Liriodendron*
Hemlock	*Tsuga*	Tupelo	*Nyssa*
Hickory	*Carya*	Viburnum	*Viburnum*
Honeylocust	*Gleditsia*	Wahoo	*Euonymus*
Honeysuckle	*Lonicera*	Walnut	*Juglans*
Hoptree	*Ptelea*	Waterlocust	*Gleditsia*
Hornbeam	*Carpinus*	Whitebeam	*Sorbus*
Horsechestnut	*Aesculus*	Whitecedar	*Chamaecyparis/Thuja*
Ironwood	*Ostrya*	Willow	*Salix*
Japanese Pagodatree	*Sophora*	Witchhazel	*Hamamelis*
Japanese Zelkova	*Zelkova*	Yellowwood	*Cladrastis*
Juniper	*Juniperus*	Yew	*Taxus*

WISCONSIN'S CHAMPION TREES

Abies concolor (White Fir)
Monroe
Green County

County	Location	CBH	H	S	Last Measured	Nominator

Abies balsamea   Balsam Fir

County	Location	CBH	H	S	Last Measured	Nominator
Iron	Mercer- T42N R2E Sec. 1 NW NE, behind Prag Residence 1/4 mi W of Town Dump	72	62	25	Donald R. Peterson (1988)	Edwin Prag (1988)
Oneida	Town of Lynn - T36N R4E Sec. 7 SW SE	66	80	24	Tim Friedrich (1987)	Tim Friedrich (1987)
Price	Georgetown - T36N R2W Sec. 14 NE SE	66	65	16	Richard Windmoeller (1983)	Richard Windmoeller (1983)
Taylor	Medford - T32N R1W Sec. 4 NW1/4 NW1/4	60	78	26	Michael J Riegert (1992)	Herbert Messman (1992)

Abies concolor   White Fir

County	Location	CBH	H	S	Last Measured	Nominator
Green	Monroe- 1907 29th Ave, T01N R07E Sec. 2 NE NE, in front of cemetery office N42° 35.588' W89° 37.485'	109	75	50	R. Bruce Allison (2004)	Ray Amiel (1988) 94" CBH
Waukesha	Delafield- 1525 Weber Court backyard	88	70	35	Joe Dietrich (1985)	Joe Dietrich (1985)

Acer campestre   Hedge Maple

County	Location	CBH	H	S	Last Measured	Nominator
Milwaukee	Wauwatosa- 7733 Stickney Ave backyard	104	82	58	Loretta Hernday (1992)	Loretta Hernday (1992)

County	Location	CBH	H	S	Last Measured	Nominator

Acer negundo   Boxelder

County	Location	CBH	H	S	Last Measured	Nominator
Manitowoc	Coopertown Township- Maribel Fire No. 171, Keeman Rd	162	40	66	P J Holschbach (1980)	P J Holschbach (1980)
Milwuakee	Whitefish Bay- 5123 Idlewild Ave back-yard, along S lot line by driveway	132	40	56	Julio Rivera (2002)	Julio Rivera (2002)

Acer nigrum   Black Maple

County	Location	CBH	H	S	Last Measured	Nominator
Washington	Town of Erin- T9N R18E Sec. 8 NW SW on line fence	152	69	61	H. Wachsmith	H. Wachsmith
Dane	Maple Bluff- 717 Lakewood Blvd side yard by house N43° 06.943' W089° 22.351'	135	80	75	R. Bruce Allison (2004)	G.H. Malek (2004)
Green	Town of Albany- T13N R3E Sec. 31 SW SW	116	94	64	Ray Amiel (1987)	Ray Amiel (1987)

Acer palmatum   Bloodleaf Japanese Maple

County	Location	CBH	H	S	Last Measured	Nominator
Brown	Green Bay- 615 Pine Terrace	26	29	21	M. Freberg (2004)	O. Monfils (1995) 16" CBH

Acer nigum (Black Maple)
Maple Bluff
Dane County

Acer platanoides (Norway Maple)
Coloma
Waushara County

County	Location	CBH	H	S	Last Measured	Nominator

Acer platanoides · Norway Maple

County	Location	CBH	H	S	Last Measured	Nominator
Waushara	Coloma- N17 10 Semrow St	179	72	70	B-W Hoffmann (2004)	(1982)
Grant	Fennimore- 680 10th St	144			Ruben Maver (1979)	Ruben Maver (1979)
Waukesha	Menomonee Falls- Mader Farmhouse N96 W15009 County Line Rd on S side of street NE corner of yard	137	74	73	Dawn & Bob Krause (1996)	Dawn M Krause (1996)

Acer pseudoplatanus · Sycamore Maple

County	Location	CBH	H	S	Last Measured	Nominator
Milwaukee	Milwaukee- 8661 N 76th Pl, St.Catherine's Place	99	41	39	(1986)	(1986)
Milwaukee	Milwaukee- E of 61 SE Garfield Reservoir Park	91	48	31	R. Rideout (1983)	R. Rideout (1983)
Milwaukee	Milwaukee- E of 61 SE Garfield Reservoir Park	76	48	34	R. Rideout (1983)	R. Rideout (1983)
Brown	Green Bay- 902 S Madison (Porlier side)	63	65	28	H. Plansky (2004)	T. Lang (1994) 59" CBH

Acer rubrum · Red Maple

County	Location	CBH	H	S	Last Measured	Nominator
Brown	Howard- Meadowbrook Park 50 yards E. of shelter	175	62	68	D. Hartman (2004)	Devon Schoening (1996) 169" CBH
Clark	Abbotsford- STH 29 W, 2-3 Bl W Jnc STH 13 & 29	162	75	76	Michael Riegert (1989)	Michael Riegert (1989)
Waupaca	Town of Royalton- T22N R13E Sec. 24 NE SE, near White Lake Rd., S side of Hwy X	162	75	60	Michael Bednarek (1988)	Michael Bednarek (1988)

County	Location	CBH	H	S	Last Measured	Nominator

Acer saccharinum — Silver Maple

County	Location	CBH	H	S	Last Measured	Nominator
Columbia	Town of Marcellon- T13N R10E Sec. 22 NW SW, N of Pardeeville, 1/4 mi E of Hwy 22, 1/4 mi N of Military Rd.	293	80	110	Michael Bednarek (1988)	Michael Bednarek (1988)
La Crosse	Mindoro- W3981 Co. D, 300' E on CTH D from jnc of Hwy 108 & Co. D	266	96	94	James E. Melton (1991)	Sandra Asleson (1991)
LaCrosse	Town of Hamilton- T16N R6W Sec 9, .1 mi E of Hwy M on S side Pleasant Valley Rd	264	73	94	Robert G. Machotka (1984)	Robert G. Machotka (1984)
Laffeyette	Wayne Township- T1N R5E Sec. 10 SE SE	253	73	70	Julie Miller (1986)	Emil Maurer (1986)

Acer saccharum — Sugar Maple

County	Location	CBH	H	S	Last Measured	Nominator
Jefferson	Town of Sullivan- N4642 Highland Drive, on walking trl near property line, Weidner property, N43° 00.259' W088° 33.380'	206	76	69	R. Bruce Allison (2004)	Mrs. A. Weidner (1974)
Waushara	Town of Wautoma- T19N R10E Sec. 20 NE SE, on Blackhawk Rd.	160	61	70	Michael Bednarek & Milo Olson (1981)	Michael Bednarek & Milo Olson (1981)
Kenosha	Town of Somers- T2N R22E NE SE	158	70	79	Michael J. Schneider (1988)	Michael J. Schneider (1988)
Kewaunee	Kewaunee- N4579 CTH B, T23N R25E Sec. 04, NE side of house	156	63	73	Phyllis Doperzlski (2002)	JW Doperzlski (2002)

Acer saccharum (Sugar Maple)
Sullivan
Jefferson County

County	Location	CBH	H	S	Last Measured	Nominator

Aesculus glabra — Ohio Buckeye

County	Location	CBH	H	S	Last Measured	Nominator
Walworth	Lake Geneva- 930 Bayview Dr. located at one end of tennis court	178			E. Hasselkus (1981)	E. Hasselkus (1981)
Milwaukee	Milwaukee- 1831 E Euclid S of intersection of Kinnickinnic & Oklahoma	168	75	50	R. Bruce Allison (2004)	Katherine & Michael Dinauer (1987)
Brown	Green Bay- 822 Grignon St. rear yard of Odd Fellows home	159	57	55	M. Freberg & E. Muecke (2004)	J F Reynolds & Tim Lang (1969) 138" CBH
Brown	Allouez- 1542 Webster Ave, Woodlawn Cemetery	147	66	56	B. Lange (2004)	T. Lang (1994) 134" CBH

Aesculus hippocastanum — Common Horsechestnut

County	Location	CBH	H	S	Last Measured	Nominator
Winnebago	Oshkosh- 717 W South Park Ave.	151	70	54	M. Bednarek (1988)	M. Bednarek (1988)
Ozaukee	Mequon- 12340 N Granville Rd.	138	42	63	G. Slusser (1973)	G. Slusser (1973)
Kenosha	Kenosha- 121 66th St.	134	65	55	M. Schneider (1989)	M. Schneider (1989)

Aesculus glabra (Ohio Buckeye)
Milwaukee
Milwaukee County

Aesculus flava (Yellow Buckeye)
Verona
Dane County

County	Location	CBH	H	S	Last Measured	Nominator

Aesculus flava — Yellow Buckeye

County	Location	CBH	H	S	Last Measured	Nominator
Richland	Sylvan Township-Sec. 10 SE1/4 SE1/4	114	72	48	(1984)	(1984)
Dane	Verona-209 S. Franklin St. front yard	93	55	60	R. Bruce Allison (2004)	S. Van Akreren (1983) 76" CBH
Columbia	Town of Dekorra-T11N R9E Sec. 25 in W part of Mackenzie Environmental Center Arboretum W of sidewalk & N of office	69	57	34	Derrick Duane (2004)	Kenneth Wood (1985) 47" CBH
Milwaukee	Milwaukee-3312 N Lake Dr. at entrance gate	54	46	37	E. Hasselkus & R. Rideout (1986)	E. Hasselkus & R. Rideout (1986)

Aesculus x carnea — Red Horsechestnut

County	Location	CBH	H	S	Last Measured	Nominator
Milwaukee	Milwaukee-3288 N Lake Dr.	59	38	27	(1986)	E. Hasselkus (1976)

Ailanthus altissima — Tree of Heaven

County	Location	CBH	H	S	Last Measured	Nominator
Milwaukee	Milwaukee-1610 S. 32nd St.	109	63	48	V. Kringer (1983)	V. Kringer (1983)
Milwaukee	Milwaukee-2602 S Superior St.	105	45	46	R. B. Rideout (1983)	R. B. Rideout (1983)
Milwaukee	Milwaukee-3811 W State St.	100	52	54	R. B. Rideout (1983)	R. B. Rideout (1983)
Milwaukee	Milwaukee-2715 Mckinley Blvd backyard	100	45	51	Kim M Gorenc (1993)	Jim Uhrinak (1993)

County	Location	CBH	H	S	Last Measured	Nominator

Alnus glutinosa   European Black Alder

County	Location	CBH	H	S	Last Measured	Nominator
Waukesha	Pewaukee- N30 W26488 Peterson Dr.	94	56	35	S. Binnie (1992)	S. Binnie (1992)
Walworth	Delavan- Terrace Park	67			M. Schneider (1981)	M. Schneider (1981)
Grant	Plattville- 390 W. Pine St.	66			C. Tiedemann (1973)	C. Tiedemann (1973)

Alnus rugosa   Speckled Alder

County	Location	CBH	H	S	Last Measured	Nominator
Brown	Suamico- Barkhausen Waterfowl Preserve 2024 Lakeview Drive	29	33	26	(1995)	Douglas R. Hartman (1986)
Sheboygan	Town of Wilson- T14N R23E Sec. 14 NE NW, approx 350' E of Black River & approx 1500' S of N line	21	25	20	(1980)	Cliff Germain (1967)

Amelanchier laevis   Allegany Serviceberry

County	Location	CBH	H	S	Last Measured	Nominator
Jefferson	Town of Hebron- T6N R15E Sec. 30 SE SW	46	51	26	Bob Van Pelt (1984)	Ed Hasselkus
Brown	Green Bay- 1022 Ridge Rd	15	28	18	M. Freberg & E. Muecke (2004)	T. Lang (1994)

County	Location	CBH	H	S	Last Measured	Nominator

Betula alleghaniensis   Yellow Birch

County	Location	CBH	H	S	Last Measured	Nominator
Iron	Knight- T45N R1E Sec. 35 SE SE	161	38	71	Rudy Kangas (1990)	Rudy Kangas (1990)
Lincoln	Tomahawk- W8397 HWY CC, T44N R2E Sec. 28 NE NW	152	90	45	Kevin M. Ires (2000)	Kevin M. Ires (2000)
Ashland	Town of Jacobs- T43N R1W Sec. 4 SE NE, at camp "E" Rd & trl # 11 go W 3/4 mi, N on woods rd 792' W, 330' b/w 2 wet areas	134	87	56	Tom Piikkila (1996)	Tom Piikkila (1996)

Betula nigra   River Birch

County	Location	CBH	H	S	Last Measured	Nominator
Vernon	Town of Bergen- T14N R7E Sec. 4, on old rd, Goose Island approx 1/2 mi S of gate	136	63	90	James Melton	James Melton
Sauk	Baraboo- Aldo Leopold Conservation Area, 8 mi N of Baraboo on Hwy T, 2-3 mi E on Levy Rd. N side of Leopold's shack	108	45	50	John Sauer & Chris Goodwin (2004)	Rodger Schley (1999) 104" CBH
Crawford	Prairie du Chien- 821 N Main St. by river in backyard near silver wind vane	99	65	68	Jeff DuCharme (2002)	Jeff DuCharme (2002)

Betula papyrifera (Paper Birch)
Woodruff
Vilas County

County	Location	CBH	H	S	Last Measured	Nominator

Betula papyrifera Paper Birch

County	Location	CBH	H	S	Last Measured	Nominator
Milwaukee	Milwaukee- 525 W. Brentwood Ln	114	58	58	William E. Husting (1979)	William E. Husting (1979)
Vilas	Woodruff- 8770 HWY J, T39N R07E Sec. 12 SW NE	94	100	53	B-W Hoffmann (2004)	Todd Anderson (2002)

Betula pendula European White Birch

County	Location	CBH	H	S	Last Measured	Nominator
Brown	Town of Humbolt- Cemetery Rd, CTH N & Ronsman Rd	103	58	60	T. Lang (1994)	Peggy Ward (1994)
Kenosha	Kenosha- 7713 34th Ave	99			(1983)	(1983)

Betula populifolia Gray Birch

County	Location	CBH	H	S	Last Measured	Nominator
Washington	Hartford- 6000 HWY 60 E	66	37	48	Bob Hults (2004)	Bob Hults (2004)
Trempealeau	Galesville- N23378 Berninski Ln; T19N R09W Sec. 22 SW SE, Kastes Morningside Orchard, walk NW from orchard buildings	46	64	29	Cindy Casey (2001)	Cindy Casey (2001)

Carpinus caroliniana American Hornbeam

County	Location	CBH	H	S	Last Measured	Nominator
Taylor	Medford- 618 E Broadway backyard	39	19	26	Michael Riegert (1989)	Michael Riegert (1989)

County	Location	CBH	H	S	Last Measured	Nominator

Carya cordiformis   Bitternut Hickory

County	Location	CBH	H	S	Last Measured	Nominator
Walworth	Lake Geneva- 1540 Evergreen Ln	120	66	87	Rudolph F. Lange (1981)	Rudolph F. Lange (1981)
Milwaukee	Milwaukee- 3614 Humbolt Blvd, Kern Park	92	90	80	D. Garacci (1983)	D. Garacci (1983)
Calumet	Woodville- T20N R19E Sec. 23 SE SE	86	117	57	Norbert & Harry Schwabenlander (1990)	Norbert & Harry Schwabenlander (1990)

Carya glabra   Pignut Hickory

County	Location	CBH	H	S	Last Measured	Nominator
Washington	Richfield- Zimmers' residence on SW side of lake AmyBelle	112	82	76	Bob Hults (2004)	Bob Hults (2004)
Milwaukee	Milwaukee- 2405 W Forest Home Ave, Forest Home Cemetery near Netzon Stone	61			S. E. Roesch (1983)	S. E. Roesch (1983)

Carya illinoinensis   Pecan

County	Location	CBH	H	S	Last Measured	Nominator
Walworth	Lake Geneva- 5N Lakeshore Dr. near lake	57			John K Raup (1981)	John K Raup (1978)
Crawford	Prairie du Chien- Catherine Wilkinson Property	44			R. D. Campbell (1979)	R. D. Campbell (1979)

Carya lacinosa   Shellbark Hickory

County	Location	CBH	H	S	Last Measured	Nominator
Dane	Town of Montrose- Sec. 30	64	46	50	(1990)	Mrs. Robert Pauli (1984)
Grant	Platteville- 685 Jewett St	49			C. Tiedemann (1973)	C. Tiedemann (1973)
Grant	Platteville- 685 Jewett St	37			C. Tiedemann (1973)	C. Tiedemann (1973)

County	Location	CBH	H	S	Last Measured	Nominator

Carya ovata — Shagbark Hickory

County	Location	CBH	H	S	Last Measured	Nominator
Brown	Green Bay- 822 Grignon	115	77	54	M. Freberg & E. Muecke (2004)	T Lang (1994) 104" CBH
Brown	Green Bay- 2420 Nicolet Dr, Shorewood Golf Course, 2nd Green	111	68	72	M. Freberg & E. Muecke (2004)	M. Freberg (1994) 106" CBH

Castanea dentata — American Chestnut

County	Location	CBH	H	S	Last Measured	Nominator
Trempealeau	Galesville- W20868 South St., T18N R8W Sec. 29 SW NW follow	169	79	50	Cindy Casey (2001)	Cindy Casey (2001)
Trempealeau	Tremplelau- Phillip Lunde Farm	139	74	58	J & S Davis (1981)	J & S Davis (1981)
LaCrosse	Town of Hamilton- T17N R6W Sec. 15 SE NE	136	76	68	James Melton (1983)	James Melton (1975)

Castanea mollissima — Chinese Chestnut

County	Location	CBH	H	S	Last Measured	Nominator
Waukesha	Chenequa- 5599 N. Hwy.83	32	41	38	Jeff Kante & Don Lund (1990)	Jeff Kante & Don Lund (1990)

Catalpa speciosa — Northern Catalpa

County	Location	CBH	H	S	Last Measured	Nominator
Grant	Platteville- 1 University Plaza, 30 yds S of Karrmann library UW-Platteville campus	250	66	70	Jean Baker-Thornton (2004)	Mike Udelhofen (2002) 237" CBH
Green	Town of Monroe- N3150 HWY 81, Pleasant View Complex, T2N R7E Sec. 27, N42° 37.383' W89° 39.428'	209	85	75	R. Bruce Allison (2004)	Ray Amiel (1986) 168" CBH

Catalpa speciosa (Northern Catalpa)
Platteville
Grant County

Catalpa speciosa (Northern Catalpa)
Monroe
Green County

County	Location	CBH	H	S	Last Measured	Nominator

Celtis occidentalis   Common Hackberry

County	Location	CBH	H	S	Last Measured	Nominator
Dane	Town of Albion-T5N R12E Sec. 13	212	76	74	(1973)	Ken P. Robert (1962)
Sauk	Baraboo- E10627 CTH W, T11N R6E Sec. 4 SE SW, 1.5 mi W of HWY 12/CTH W intrsct	155	94	64	Chris Goodwin & John Sauer (2004)	Rick Livingston (2001) 152" CBH
Jefferson	Lake Mills- W6480 Hwy A, 3 mi E of Lake Mills on Highway A	152	89	77	John C Stein (1991)	John C Stein (1991)
LaCrosse	LaCrosse- 123 S. 11th St backyard	128	63	63	Elizabeth Ash (1986)	Elizabeth Ash (1986)

Cercidiphyllum japonicum   Katsuratree

County	Location	CBH	H	S	Last Measured	Nominator
Milwaukee	Milwaukee- 207 Lake Dr., Wil-O-Way Grant Park behind Wil-O-Way Center S of wading pool N42° 55.513' W87° 50.925'	130	61	54	R. Bruce Allison (2004)	R. Hodkiewicz (1978) 61" CBH
Walworth	Lake Geneva- W2765 S. Lake Shore Dr, formerly McNally house 850 S. Lakeshore Drive, N42° 34.490' W88° 26.242'	115	60	50	R. Bruce Allison (2004)	(1981) 99" CBH
Dane	Madison- 4110 Mandan Crecent	60	65	40	R. Bruce Allison (2004)	M. A. Boettger (1973)

Cercidiphyllum japonicum (Katsuratree)
Lake Geneva
Walworth County

Cercis canadensis (Eastern Redbud)
Madison
Dane County

County	Location	CBH	H	S	Last Measured	Nominator

Cercis canadensis — Eastern Redbud

County	Location	CBH	H	S	Last Measured	Nominator
LaCrosse	LaCrosse-2102 Winnebago St.	59	27	43	Jay Fernholz (1999)	Jay Fernholz (1999)
Brown	Green Bay-344 Gray	47	24	29	M. Freberg & E. Muecke (2004)	T Lang (1994) 59" CBH
Dane	Madison-826 Minakwa Drive, side yard	42	28	35	R. Bruce Allison (2004)	Sue Krause (2004)

Chamaecyparis pisifera — Sawara Falsecypress

County	Location	CBH	H	S	Last Measured	Nominator
Kenosha	Kenosha-4426 5th Ave	42	35	20	(1989)	(1989)
Milwaukee	Milwaukee-3062 S Superior	36	30	10	S. F. Roesch (1982)	S. F. Roesch (1982)
Milwaukee	Milwaukee-3288 N Lake Dr	26	26	16	E. Hasselkus & R. Rideout (1986)	E. Hasselkus & R. Rideout (1986)

Cladrastis lutea — American Yellowwood

County	Location	CBH	H	S	Last Measured	Nominator
Walworth	Williams Bay-Yerkes Observatory	118	55	45	Ed Struble (2004)	E. R. Hasselkus (1976) 77" CBH
Milwaukee	Milwaukee-2316 E Edgewood	62	42	39	(1983)	Otting & Hasselkus (1976)

Cornus alternifolia — Pagoda Dogwood

County	Location	CBH	H	S	Last Measured	Nominator
Kenosha	Paris Township- Sec. 18 SE 1/4 NE 1/4, N side of Hwy 142 about 1/4 mi W	23	14	25	M. Schneider & Phil Sander (1987)	M. Schneider & Phil Sander (1987)

Craetegus mollis (Downy Hawthorne)
Madison
Dane County

Craetegus punctata (**Dotted Hawthorne**)
Maple Bluff
Dane County

38

County	Location	CBH	H	S	Last Measured	Nominator

Corylus colurna   Turkish Filbert

| Brown | Green Bay- 715 S. Oakland | 30 | 31 | 28 | T. Lang (1994) | T. Lang (1994) |

Cotinus obovatus   American Smoketree

| Brown | Green Bay- 516 S. Webster | 66 | 55 | 45 | T. Barber (2004) | T. Lang & M. Freberg (1994) 59" CBH |
| Oconto | Gillett- 1st & Birch Sts., 2nd house E of S side of St | 63 | | | L. Krause (1986) | L. Krause (1986) |

Craetegus mollis   Downy Hawthorn

| Dane | Madison- UW Campus Red Gym, 716 Langdon, UW ID# 69-7346 | 72 | 30 | 45 | Daniel Einstein & Stuart Sonnedecker (2004) | B. Thomas (1975) |
| Brown | Allouez- 3325 Webster Ave. | 68 | 32 | 43 | P. Rudquist (1994) | R. Nierdaels (1994) |

Craetegus punctata   Dotted Hawthorn

| Dane | Maple Bluff- 22 Fuller Drive-side yard along road, N43° 05.932' W089° 22.157' | 74 | 25 | 35 | R. Bruce Allison (2004) | M. Schneider (1980) 58" CBH |

Elaeagnus angustifolia   Russianolive

| Brown | Green Bay- 1273 Dousman | 114 @ 2' | 47 | 53 | M. Freberg & E. Muecke (2004) | T. Lang (1994) 97" CBH |
| Brown | Town of Scott- Wequiock Falls | 106 @ 2' | 40 | 39 | M. Freberg & E. Muecke (2004) | M. Freberg (1994) 104" CBH |

County	Location	CBH	H	S	Last Measured	Nominator

Euonymus alatus — Winged Euonymus

County	Location	CBH	H	S	Last Measured	Nominator
Dane	Madison- UW Campus, SE corner of old Music Hall, UW ID# 68-954	26	20	30	R. Bruce Allison (2004)	Boettger Wood (1973)
Brown	Green Bay- 1112 Doty St	25	17	18	H. Plansky (2004)	P. Hartman (1994) 16" CBH

Euonymus atropurpureus — Eastern Wahoo

County	Location	CBH	H	S	Last Measured	Nominator
Milwaukee	Milwaukee- Jackson Park, near bridge over stream along 35th St	132			S. E. Roesch (1983)	S. E. Roesch (1983)

Euonymus europaea — Spindletree Euonymus

County	Location	CBH	H	S	Last Measured	Nominator
Kenosha	Kenosha- Simmons Library, 711 59th Place	34	25	33	M. Schneider (1988)	M. Schneider (1988)
Brown	DePere- St. Norbert College, just W of power plant	24	31	28	D. Melichar (2004)	P. J. Holschbach & J. Landwehr (1980) 24" CBH
LaCrosse	LaCrosse- 2700 Cass St	24	23	19	John F Zoero (1981)	John F Zoero (1981)

Euonymus hamiltoniana — Yeddo Euonymus

County	Location	CBH	H	S	Last Measured	Nominator
Brown	Green Bay- 3270 Nautical Court	55	15	29	T. Barber (2004)	G. Spevacek (1982) 49" CBH

Euonymus alatus (**Winged Euonymus**)
Madison
Dane County

County	Location	CBH	H	S	Last Measured	Nominator

Fagus grandifolia   American Beech

County	Location	CBH	H	S	Last Measured	Nominator
Kewaunee	Town of Casco- T24N R24E Sec. 34, near campsite 21	132	82	77	Sue Matuszewski (1989)	Sue Matuszewski (1989)
Milwaukee	Milwaukee- 3257 S. Lake Dr., located in seminary woods 24 ft of post #5	112	85	60	Sandy Sue Szanderek (1984)	Sandy Sue Szanderek (1984)
Manitowoc	Town of Cato- T19N R22E Sec. 9 NW 1/4 SE SE	108	100	58	Norbert Pritzl (1989)	Norbert Pritzl (1989)
Dodge	Town of Williamstown- T12N R16E Sec. 36 SW SW, tree near woods trail on N side of 40	108	86	65	J. Malesevich & A. Magyar (1986)	J. Malesevich & A. Magyar (1986)

Fagus sylvatica   European Beech

County	Location	CBH	H	S	Last Measured	Nominator
Milwaukee	Milwaukee- South Shore Park, Estes St	157	51	56	(1988)	D. Rozmarynoski
Milwaukee	Milwaukee- Lake Park, off Lake Dr. near tennis courts	142	75	57	B-W Hoffmann (2004)	D. Konieczka (1978)
Milwaukee	Milwaukee- 2216 East Estes	141	58	62	(1983)	(1983)
Milwaukee	Milwaukee- Washington Park, N of basketball court, E of asphalt walk.	133	46	50	J. Kringer (1983)	J. Kringer (1983)

Fagus sylvatica (European Beech)
Lake Park, Milwaukee
Milwaukee County

County	Location	CBH	H	S	Last Measured	Nominator

Fraxinus americana — White Ash

County	Location	CBH	H	S	Last Measured	Nominator
Kenosha	Brighton Township-Bong Recreation Area, 42° 38.70'N 88° 07.797'W, T2N R20E Sec. 10 NW1/4 SW1/4 on E side of access road	181	35	40	Bill Manke (2002)	W. E. Scott (1963), A. Westerman & Susan Hill (1983)
Brown	Howard-3215 Shawano	168	70	55	D. Hartman (2004)	H. Plansky (1994) 154" CBH
Dane	Madison- 100 yds N of state of Wis. Property, 100 yds E of Lake Mendota Dr.	157	90	82	Bob Van Pelt (1985)	Bob Van Pelt (1985)
Manitowoc	Town of Newton-T18N R23E Sec. 10	149	77	86	Ralph Pleuss (1995)	Ralph Pleuss (1991)

Fraxinus excelsior — European Ash

County	Location	CBH	H	S	Last Measured	Nominator
Brown	Green Bay-1215 Lark	88	43	50	M. Freberg & E. Muecke (2004)	M. Freberg (1994) 62" CBH
Walworth	Whitewater- Chopp Arboretum, UW Whitewater, in front of Old Main	88	40	45	Steven Bertagnolli (2004)	R. Miner (1982) 46" CBH
Brown	Green Bay-1293 Lark	84	44	47	M. Freberg & E. Muecke (2004)	T. Lang (1994) 58" CBH
Milwaukee	Milwaukee-Greenwood Cemetery, E of building inside fence along Cleveland Ave.	62			Hasselkus (1980)	Hasselkus (1980)

County	Location	CBH	H	S	Last Measured	Nominator

Fraxinus nigra — Black Ash

County	Location	CBH	H	S	Last Measured	Nominator
Sawyer	Town of Radisson- T39N R7W Sec. 36 SW SW	102	90	43	P. Agurkis M. Fries J. Benson (1990)	P. Agurkis M. Fries J. Benson (1990)
Iron	Knight- T45N R1E Sec. 36 SE SW in timber sale in Iron County forest	101	88	40	Gary Glonek & Charles Zinsma (1990)	Gary Glonek & Charles Zinsma (1990)
Sawyer	Winter- T37N R3W Sec. 30 SW1/4 NE 1/4 SE 1/4, 4.75 mi Hwy M/Bear Creek Rd. intersect. Travel 35 yds north on Bear Creek Rd.	90	94	35	Donald Hoeft & Susanne Brown (1996)	Donald Hoeft & Susanne Brown (1996)
Douglas	Town of Summit- T45N R14W Sec. 26 SW NE	87	93	38	Gary Zileske (1984)	Gary Zileske (1984)

Fraxinus pennsylvanica — Green Ash

County	Location	CBH	H	S	Last Measured	Nominator
Rock	Newville- 10975 Hillside Rd. in yard	224	53	73	Mary Ann Kroehn Buenzow & Chris Ranum (2004)	Thomas Rausch
Dodge	Waupun- Hwy 49 E side of Waupun on Main St/Hwy 49. Tree b/w Hwy 26/Watertown St & Grove St	148	53	36	Rodger Schley (2000)	Rodger Schley (2000)
Green	Town of Albany- T3N R9E Sec. 4 SW SW in middle of red pine plantation	145	73	85	Ray Amiel (1988)	Ray Amiel (1988)
Waukesha	Oconomowoc- 659 E Juneau Ave backyard	144	36	36	Mrs. Alex Henschel (1980)	Mrs. Alex Henschel (1980)
Dane	Verona- 157 Paoli St.	133	60	75	R. Bruce Allison (2004)	C. W. Handrich (1981) 114" CBH

Fraxinus pennsylvanica (Green Ash)
Verona
Dane County

Ginkgo biloba (Ginkgo)
Monroe
Green County

Ginkgo biloba (Ginkgo)
Lake Geneva
Walworth County

County	Location	CBH	H	S	Last Measured	Nominator

Fraxinus quadrangulata

Blue Ash

County	Location	CBH	H	S	Last Measured	Nominator
Dane	Madison- UW Campus, N of Home Economics, UW ID# 67-916	72	45	49	Daniel Einstein (2004)	B. Thomas (1975) 54" CBH
Dane	Madison- UW Arboretum in Longnecker Garden, ash collection	55	35	47	Ken Zuba (2004)	Ken Zuba (2004)

Fraxinus velutina

Modesto Ash

County	Location	CBH	H	S	Last Measured	Nominator
Milwaukee	Milwaukee- 8050 W Seranton	60	38	33	R. Thurow (1983)	R. Thurow (1983)
Milwaukee	Milwaukee- 8117 W. Custer	45	30	32	J Boeder & R Rideout (1983)	J Boeder & R Rideout (1983)
Milwaukee	Milwaukee- 8133 W Custer	39	26	25	J Boeder & R Rideout (1983)	J Boeder & R Rideout (1983)

Ginkgo biloba

Ginkgo

County	Location	CBH	H	S	Last Measured	Nominator
Green	Monroe- 1205 13th St., N42° 35.994', W89° 38.612'	192	75	85	R. Bruce Allison (2004)	J F Reynolds (1969)
Walworth	Lake Geneva- Covenant Harbor Camp, Pier 30, N42° 35.057', W88° 22.049'	166	80	71	R. Bruce Allison (2004)	Willis Erickson (1966)
Milwaukee	Milwaukee- 2128 Lafayette Plc, Intrsct w/ North Terrace	142	77	51	Adam Schmidt (2002)	Adam and Mark Schmidt (2002)
Dane	McFarland- 6009 Exchange St., front yard, N43° 00.758', W89° 17.400'	125	55	56	Bill Manke (2002)	James Fitzpatrick (1980)

County	Location	CBH	H	S	Last Measured	Nominator

Gleditsia triacanthos   Common Honeylocust

County	Location	CBH	H	S	Last Measured	Nominator
Rock	Village of Cooksville- In a fence row directly across the street (E) of 11117 N. Church St.	179	75	51	Mary Ann Kroehn Buenzow & Chris Ranum (2004)	Michael Bendarek (1988) 172" CBH
Sauk	Lakeway Resort, Corner hwy 123 & Gall Rd. 1 mi. S of Baraboo	170	80	66	Chris Goodwin & John Sauer (2004)	James Kremtreiter (1986) 151" CBH
Dane	Mazomanie- 430 Bridge, across from vacant lot b/w 409 & 417 Bridge St.	166	60	76	Carla Wolf (2002)	Carla Wolf (2002)
Grant	Prairie du Chien- T5N R6W Sec. 7 NW 1/4 SW 1/4	154	94	69	Dan Oles and Loren Danson (1995)	Dan Oles and Loren Danson (1995)

Gymnocladus dioicus   Kentucky Coffeetree

County	Location	CBH	H	S	Last Measured	Nominator
Jefferson	Town of Sumner- T5N R13E Sec. 17 SW SW, in woods on property along mowed path	143	122	72	Mary Ann Cisewski (1986)	Mary Ann Cisewski (1986)
Dodge	Waterloo- Portland Township, T1N R13E Sec. 15 SW1/4, E bank of Crawfish River	112	55	61	(1982)	Bill Tans (1976)
Green	Monroe- 718 22nd Ave., street tree, N42° 36.266', W89° 37.960'	96	65	55	R. Bruce Allison (2004)	Steven Miller (1980) 83" CBH

Gymnocladus dioicus (Kentucky Coffeetree)
Monroe
Green County

County	Location	CBH	H	S	Last Measured	Nominator

Halesia carolina   Silverbell

County	Location	CBH	H	S	Last Measured	Nominator
Milwaukee	Milwaukee- 2405 W. Forest Home Ave, 110 ft W of Pond & 50 ft S of "Hertzfeld" Monument. N42° 59.896, W087° 56.536	33	60	30	Dale R. Konieczka (2004)	Dale R. Konieczka (1980) 24" CBH

Juglans ailanthifolia   Heartnut

County	Location	CBH	H	S	Last Measured	Nominator
Kenosha	Paris Township- 12908 Burlington Rd, T2N R21E Sec. 24 NW1/4, Hwy 142	122	35	65	(1990)	Mike Schneider (1987)

Juglans cinerea   Butternut

County	Location	CBH	H	S	Last Measured	Nominator
Dane	Madison- 933 N Fair Oaks Ave	154	65	60	R. Bruce Allison (2004)	Celia White (2004)
Vernon	Town of Hilsboro Sec. 30 on gravel trail across Hwy 33 from farm house on N side of trail	150	63	68	James Dalton (1990)	James Dalton (1990)
La Crosse	Coon Valley- R R 1 Brinkman Ridge Road	144	80	65	Jay Fernholz (2003)	Jay Fernholz (1991) 119" CBH
Manitowoc	Manitowoc- 715 New York Ave. in yard	134	50	76	Ralph Pleuss (1989)	Ralph Pleuss (1989)

County	Location	CBH	H	S	Last Measured	Nominator

Juglans nigra

Eastern Black Walnut

County	Location	CBH	H	S	Last Measured	Nominator
Dane	Belleville- 7741 State Rd HWY 69/92, T5N R8E Sec. 31 SW SE, Driveway S off Hwy 69	189	91	83	R. Bruce Allison (2004)	Gael Dilthey (1980) R. Livingstone
Trempealeau	Village of Trempealeau, Corner of Main and Third Street	168	96	91	Scott Laurie (2002)	Al Bagley (2002)
Winnebago	Omro- 7984 Tritt Road 200' NNW of owner's house	157	74	88	John Slocum (1987)	John Slocum (1987)

Juglans regia

Carpathian Walnut

County	Location	CBH	H	S	Last Measured	Nominator
Milwaukee	Milwaukee- 4210 West Roosevelt	66	56	50	(1983)	(1983)
Milwaukee	Hales Corners- Boerner Botanical Gardens, N of parking lot near center	34	--		R. Hodkiewicz (1978)	Milton Kral (1972)

Juniperus virginiana

Eastern Red Cedar

County	Location	CBH	H	S	Last Measured	Nominator
Rock	Town of Lima- Lima Center Cemetery, T4N R14E Sec. 27, between Whitewater and Milton	105	44	33	James R. Ream (1986)	James R. Ream (1986)
Kenosha	Town of Randall- 301 CTH F, T1N R19E Sec. 16 NE SE	92	48	34	William Bloom & Mike Schneider (1987)	William Bloom & Mike Schneider (1987)
Kenosha	Brighton- DNR Bong Recreation Area, 2600' N of Hwy 142 on line next to Salem School Forest	92	45	30	A. Westerman (1983)	A. Westerman (1983)
Grant	T6N R6W Sec. 22 NE SE	88	64	35	Craig Hollingsworth (1983)	Craig Hollingsworth (1983)

Juglans nigra (**Eastern Black Walnut**)
Belleville
Dane County

County	Location	CBH	H	S	Last Measured	Nominator

Larix decidua

European Larch

County	Location	CBH	H	S	Last Measured	Nominator
Green	Spring Grove-Sec. 26 SW SW	131	95	54	R Livingston (1992)	R Livingston (1992)
Grant	Lancaster- Winskill Public School Playground	128	65	59	C. Hollingsworth & P. Malovrh (1984)	C. Hollingsworth & P. Malovrh (1984)
Iowa	Mineral Point-next to historical site No. 26	118	68	59	Jim Widder (1990)	Jim Widder (1990)

Larix gmelini

Dahurian Larch

County	Location	CBH	H	S	Last Measured	Nominator
Milwaukee	Hales Corners-Whitnall Park, b/w College Ave & Park Drive W of lagoon	35			Wm. Radler & S. McCllock (1979)	Wm. Radler & S. McCllock (1979)

Larix laricina

Tamarack

County	Location	CBH	H	S	Last Measured	Nominator
Price	Town of Know- Sec. 24, 30 yds from road, W side CTH YY	121	71	36	Chad McGrath (1985)	R.F.Wendt
Dodge	Fox Lake- College Ave, front yard of Kratz & Smeedama Funeral Home	98	85	50	Rodger Schley (2000)	Rodger Schley (2000)

Liquidambar styraciflua

Sweetgum

County	Location	CBH	H	S	Last Measured	Nominator
Kenosha	Kenosha- 6347 47th Ave., backyard	26	30		S. A. Morgan (1975)	S. A. Morgan (1975)
Milwaukee	Milwaukee- Parkway NW of Jackson Park	20			S .E. Roesch (1983)	S .E. Roesch (1983)

County	Location	CBH	H	S	Last Measured	Nominator

Liriodendron tulipifera · Tuliptree

County	Location	CBH	H	S	Last Measured	Nominator
Green	Monroe- 2021 10th St.	101	91	67	R. Bruce Allison (2004)	Ray Amiel (1988)
Kenosha	Kenosha- Green Ridge Cemetery 6604 7th Ave, located on W fenceline N of the maintenance building	92	65	42	Michael Schneider (1988)	Michael Schneider (1988)
Kenosha	Bristol- 8024 200th Ave, HWY 45 in yard	73	75	33	Mike Schneider & Phil Sander (1987)	Mike Schneider & Phil Sander (1987)

Maclura pomifera · Osageorange

County	Location	CBH	H	S	Last Measured	Nominator
Racine	City Trunk KR, N side of Road, 1-2 miles W of intsct of Hwy 31 or 1/2 mi	99	40	40	M. Schneider & E. Hasselkus (1990)	M. Schneider & E. Hasselkus (1990)
Kenosha	Kenosha- 7855 22nd Ave, street tree	84	62	40	Paul R. Pulera (1984)	Paul R. Pulera (1980)

Magnolia acuminata · Cucumbertree Magnolia

County	Location	CBH	H	S	Last Measured	Nominator
Dodge	Beaver Dam- 315 Park Ave	148	66	68	Thomas J. Pyrek (1991)	Erin McCarthy (1991)
Sauk	Village of Spring Green- S edge of lot on E corner of intsct of Winsted & Jefferson	100	65	52	Chris Goodwin & John Sauer (2004)	Chris Winther (1984) 69" CBH
Racine	Racine- 26300 Washington Ave, Hwy 20, across street from Dreams End Girl Scout	82	43	39	K. Jerome (1989)	C. Winther (1980)
Racine	Burlington- 548 E State (Hwy 11), side yard	70	32	25	S. Erickson (1985)	S. Erickson (1985)

Liriodendron tulipifera (Tuliptree)
Monroe
Green County

County	Location	CBH	H	S	Last Measured	Nominator

Magnolia stellata — Star Magnolia

County	Location	CBH	H	S	Last Measured	Nominator
Dane	Madison- 21 North Prospect Ave backyard	39	25	25	R. Bruce Allison (2004)	(1970) 23" CBH
Dane	Mt. Horeb- 106 South 5th Street	12	29	22	Jeff Gorman (2004)	Jeff Gorman (2004)

Magnolia x soulangiana — Saucer Magnolia

County	Location	CBH	H	S	Last Measured	Nominator
Dane	Madison- 1922 Birge Terrace front yard, formerly 1936 University Ave.	45	25	22	R. Bruce Allison (2004)	S. McCulloch (1980) 30" CBH
Brown	DePere- 642 Grant St	41	25	20	D. Melichar (2004)	P. Holschbach (1994) 37" CBH
Dane	Madison- 134 West Gilman St	29	29	34	M Fish (1975)	M Fish (1975)

Malus baccata — Siberian Crabapple

County	Location	CBH	H	S	Last Measured	Nominator
Taylor	Medford- N3879 Hwy.13	88	33	46	(1989)	(1989)

Malus pumila — Common Apple

County	Location	CBH	H	S	Last Measured	Nominator
Sheboygan	Random Lake-W7416 Valley View Road	115@4'	21	35	Bob Hults (2004)	Bob Hults (2004)
Dodge	Theresa Township- Sec. 14 SE 1/4 intrsct Hwy "P" & 175	104			Alan C. Pape (1986)	Alan C. Pape (1986)

Malus 'Snowdrift' — Snowdrift Crabapple

County	Location	CBH	H	S	Last Measured	Nominator
Brown	Green Bay- 1683 9th St	60	27	40	M. Freberg & E. Muecke (2004)	T. Lang (1994) 44" CBH

Magnolia x soulangiana (Saucer Magnolia)
Madison
Dane County

County	Location	CBH	H	S	Last Measured	Nominator

Metasequoia glyptostroboides   Dawnredwood

County	Location	CBH	H	S	Last Measured	Nominator
Milwaukee	Brown Deer- 8482 N 52nd St.	45	46	24	J DiFrances (1991)	J DiFrances (1991)
Kenosha	Kenosha- 6501 3rd Ave, Kemper Center, S of Chapel entrance	43	30	33	M Schneider (1988)	M Schneider (1988)
Milwaukee	Milwaukee- 3526 S Pine	42	37	26	B. Kechaver & R. Rideout (1983)	B. Kechaver & R. Rideout (1983)

Morus alba   Russian Mulberry

County	Location	CBH	H	S	Last Measured	Nominator
Washington	Germantown- W188- N11881 Maple Rd., NE1/4 of Sec. 20.	201@2'	27	55	Al Johnson (2004)	Unknown
Winnebago	Oshkosh- 2819 Waupun Rd	138			John Green (1979)	John Green (1974)
Kenosha	Brighton- T2N R20E Sec. 10	134	35	50	M. Schneider (1983)	Walter E. Scott (1963)

Morus rubra   Red Mulberry

County	Location	CBH	H	S	Last Measured	Nominator
Walworth	Elkhorn- N7019 Hwy 12 & 67, 1 mi N of HWY A, 100 yds due W of Sugar Creek Pantry parking lot. 5 mi N of Elkhorn	128	48	43	Ken Clark (2001)	Ken Clark (2001)
Buffalo	Town of Nelson- T22N R13W SEC. 8 NE SE	112	45	40	(1983)	Edin D. Godel (1970)
Calumet	Hilbert- N6985 Irish Rd., E of house	105	57	10	M. Spreeman & B. Hipp (2002)	M. Spreeman (2002)

County	Location	CBH	H	S	Last Measured	Nominator

Nyssa sylvatica · Black Tupelo

County	Location	CBH	H	S	Last Measured	Nominator
Dane	Madison- UW Arboretum at the McKay Center	49	25	25	R. Bruce Allison (2004)	E. Hasselkus (2004)
Kenosha	Berryville- 6th Place: 1/2 block W of Hwy 32, first street N of County A	29			Boettger-Wood (1978)	Boettger-Wood (1973)
Milwaukee	Milwaukee- 3521 Prospect Ave	27	36	20	E. Hasselkus & R. Rideout (1986)	E. Hasselkus & R. Rideout (1986)

Ostrya virginiana · Ironwood

County	Location	CBH	H	S	Last Measured	Nominator
Brown	Green Bay- 338 S. Quincy	70	50	41	H. Plansky (2004)	T. Lang (1994) 65" CBH
Milwaukee	Milwaukee- 201 Lake Dr., Grant Park, 300 yds E of 1st Parking lot next to a street	64			S. E. Roesch (1983)	S. E. Roesch (1983)
Manitowoc	Manitowoc- 1116 N 8th St. back yard	62	53	51	Phil Knier (2000)	Ann Unertl (2000)

Phellodendron amurense · Amur Corktree

County	Location	CBH	H	S	Last Measured	Nominator
La Crosse	Bangor- 10th Ave. South, Pleasant View Farm	153	66	66	Jay Fernholz (1993)	Jay Fernholz (1993)
Jefferson	Watertown- Western Ave., Northwestern College Campus	135	49	64	David R. Schumann (2001)	E. Hasselkus (1965)
Dodge	N side of Fox Lake. Black Hawk Trail to Maple Point at very end of road.	73	58	44	Rodger Schley (2000)	Rodger Schley (2000)
Walworth	Lake Geneva- 62 Snake Rd, Peterkin Estate	68			M. Schneider (1981)	M. Schneider (1981)

Nyssa sylvatica (Black Tupelo)
Madison
Dane County

Picea abies (**Norway Spruce**)
Shorewood Hills
Dane County

County	Location	CBH	H	S	Last Measured	Nominator

Picea abies   Norway Spruce

County	Location	CBH	H	S	Last Measured	Nominator
Dane	Shorewood Hills- 3511 Sunset Dr	156	70	65	R. Bruce Allison (2004)	Harold P. Rusch (1980) 140" CBH
Kenosha	Brighton Township- Sec.33 SW1/4 SE1/4, N side of Cty K	150	70	54	M Schneider & Phil Sander (1987)	M Schneider & Phil Sander (1987)

Picea glauca   White Spruce

County	Location	CBH	H	S	Last Measured	Nominator
Pierce	Village of Elmwood- Tree in middle of road in residential area of village on corner of May Ave. & Scott St.	140	89	25	Jamie Reitz & Josh Kern (2004)	Joanne Baier (1986)
Oneida	Town of Lynn- T36N R4E Sec. 19 NE NE, 200 yds W of S Turcot Rd off Rustic Rd	111	84	44	Glenn Witmer (1985)	Glenn Witmer (1985)
Douglas	Town of Superior- T48N R14W Sec. 27 SW SE on Nemadji River	104	109	32	Dr. Scott Nielson (1985)	Dr. Scott Nielson (1985)
Shawano	Aniwa Township- T29N R11E Sec.1 NW SW	104	84	45	Larry Forden (1991)	John Stengl (1991)

Picea mariana   Black Spruce

County	Location	CBH	H	S	Last Measured	Nominator
Taylor	Medford- W5172 Allman Ave, E bank of beaver pond on Risch land 100 yds from road	62	78	21	Nich Risch (1989)	Nich Risch (1989)
Brown	DePere- St. Norbert College	58	48	30	D. Melichar (2004)	Paul Hartman (1995) 51" CBH
Douglas	Town of Highland- T46N R10W Sec. 30 SW NE, N Shore Brule River	55	65	19	Scott Nielson (1985)	Scott Nielson (1985)

County	Location	CBH	H	S	Last Measured	Nominator

Picea omorika   Serbian Spruce

County	Location	CBH	H	S	Last Measured	Nominator
Kenosha	Kenosha- 7504 19th Ave	28	40	19	M. Schneider & E. Hasselkus (1990)	M. Schneider & E. Hasselkus (1990)
Milwaukee	Hales Corners- Boerner Botanical Gardens, 34-653 & 39-632 main conifer collection near center	27			Wm. Radler (1979)	Wm. Radler (1979)

Picea pungens   Colorado Spruce

County	Location	CBH	H	S	Last Measured	Nominator
Brown	DePere- 109 Ontario St	88	77	40	D. Melichar (2004)	Paul Hartman (1996) 81" CBH
Brown	Green Bay- 1322 Dousman	76	65	28	M. Freberg (1996)	Paul Hartman (1996)
Shawano	Town of Wittenburg- T27N R11E Sec. 17 NE NW, on S side of Hwy 29	71	71	26	(1983)	Edwin D. Godel (1961)
Shawano	Town of Wittenburg- in Wittenburg Cemetery S of Hwy M	70	61	34	Chad McGrath (1981)	Chad McGrath (1981)

Pinus banksiana   Jack Pine

County	Location	CBH	H	S	Last Measured	Nominator
Brown	Green Bay- 1018 Lyndon	57	55	29	M. Freberg & E. Muecke (2004)	T. Lang (1994) 48" CBH
Brown	Green Bay- 1713 Burns	48	42	22	M. Freberg & E. Muecke (2004)	Paul Hartman (1995) 37" CBH

County	Location	CBH	H	S	Last Measured	Nominator

Pinus cembra   Swiss Stone Pine

County	Location	CBH	H	S	Last Measured	Nominator
Milwaukee	Milwaukee- 2405 Forest Home Ave., Forest Home Cemetery, S of W Cleveland Ave. just inside fence	41			E. Hasselkus (1980)	E. Hasselkus (1980)
Brown	Allouez- 1910 S. Webster	32	34	14	M. Freberg (1994)	Paul Hartman (2004)
Kenosha	Kenosha- 6604 7th Ave., Green Ridge Cemetery	29	18	14	(1988)	M. Schneider (1984)

Pinus flexilis   Limber Pine

County	Location	CBH	H	S	Last Measured	Nominator
Columbia	Town of Dekorra- MacKenzie Environmental Center, W of Conservation Exhibit	66	60	33	Derrick Duane (2004)	Kenneth W. Wood (1985) 54" CBH
Milwaukee	Wauwatosa- Wauwatosa Cemetery	52			E. Hasselkus (1981)	E. Hasselkus (1981)

Pinus mugo   Mugo Pine

County	Location	CBH	H	S	Last Measured	Nominator
Brown	Howard- 3215 Shawano	31	25	33	D. Hartman (2004)	T. Lang (1994) 29" CBH
Dane	Madison- 1007 Hillside Ave., backyard behind pool	22	10	20	R. Bruce Allison (2004)	Terry Kelly (2004)

Pinus mugo (Mugo Pine)
Madison
Dane County

Pinus nigra (**Austrian Pine**)
Janesville
Rock County

County	Location	CBH	H	S	Last Measured	Nominator

Pinus nigra   Austrian Pine

County	Location	CBH	H	S	Last Measured	Nominator
Rock	Janesville- 2320 West Court St next to Grey Brewing Co. N42° 40.765' W89° 02.991'	167	65	80	R. Bruce Allison (2004)	(1981) 145" CBH
Walworth	Elkhorn- 302 N Wisconsin side yard	131	70	60	Boettger-Wood (1973)	Boettger-Wood (1973)
Jefferson	Fort Atkinson- Cemetery S of town on Hwy 26 near Copps Dept. Store	111	60	61	T Klitzkie (1975)	T Klitzkie (1975)
Kenosha	Parkside- UW Campus across road from heating station	101			M. Schneider (1984)	M. Schneider (1984)

Pinus ponderosa   Ponderosa Pine

County	Location	CBH	H	S	Last Measured	Nominator
Dane	Madison- 10 Babcock Dr. UW Campus, behind building, UW ID# 56-1527	85	63	32	Daniel Einstein (2004)	Bob Van Pelt (1984) 40" CBH
Brown	Green Bay- 615 Pine Terrace	60	77	26	M. Freberg & E. Muecke (2004)	M. Freberg (1995)

Pinus resinosa   Red Pine

County	Location	CBH	H	S	Last Measured	Nominator
Douglas	Town of Solon Springs- T45N R12W Sec. 25, on N edge of NE parking lot in Lucius Woods Park	102	84	45	Dr. Scott Nielson (1985)	Dr. Scott Nielson (1985)
Douglas	Town of Solon Springs- T45N R12W Sec. 25 NW SW, SW corner of parking lot in center of Lucius Woods Park	98	112.5	41	Dr. Scott Nielson (1984)	Dr. Scott Nielson (1984)
Brown	Green Bay- Fort Howard Cemetery	80	65	32	M. Freberg & E. Muecke (2004)	Paul Hartman (1978) 80" CBH

Pinus ponderosa (Ponderosa Pine)
Madison
Dane County

County	Location	CBH	H	S	Last Measured	Nominator

Pinus strobus   Eastern White Pine

County	Location	CBH	H	S	Last Measured	Nominator
Douglas	Brule River State Forest, T46N R10W Sec. 30 SW	174	119	49	Dr. Scott Nielson (1985)	Dr. Scott Nielson (1985)
Marinette	Stephenson- T33N R18E Sec. 1 NW SE about 15 mi NW of Village of Crivitz. 440 yds S of N11284 Five Star Ln	167	118	54	Dan Mertz (2000)	Dan Mertz (2000)
Douglas	Town of Gordon- T44N R13W Sec. 35 NE NW, on E bank of Arnold Creek at jnc with Moose River	166	101	51	Dr. Scott Nielson (1985)	Dr. Scott Nielson (1985)
Douglas	Town of Brule- T47N R10W Sec. 27 SW NW, on Blueberry Creek 1 mi SW of Brule ranger station	162	125	47	Dr. Scott Nielson (1984)	Dr. Scott Nielson (1984)

Pinus sylvestris   Scots Pine

County	Location	CBH	H	S	Last Measured	Nominator
Fond du Lac	Rosendale- W10335 Rose Eld Rd, NW corner of WI Hwy 26& Rose Eld Rd	170	42	48	Thaddeaus J Pyrek (1992)	P Marinak & J Zebrowski (1992)
Dane	Albion Township- Farm No 921, Sec. 28, 1/2 mile W of Albion on Bliven Rd	140	40	50	T. Klitzkie (1975)	T. Klitzkie (1975)
Jefferson	Watertown- 900 E. Cady St. front yard	138	43	56	David R Schumann (2000)	David R Schumann (2000)

County	Location	CBH	H	S	Last Measured	Nominator

Platanus occidentalis American Sycamore

County	Location	CBH	H	S	Last Measured	Nominator
Fond du Lac	Waupun- 222 N Mill St.	167	115	73	Ted Pyrek (1983)	Ted Pyrek (1983)
Fond du Lac	Fond du Lac- 100 Military Rd.	160	81	71	(1983)	Ted Pyrek (1974)
Brown	Green Bay- 1534 Morrow	132	68	77	T. Barber (2004)	Paul Hartman (1978) 125" CBH

Platanus x acerifolia London Planetree

County	Location	CBH	H	S	Last Measured	Nominator
Milwaukee	Milwaukee- Kern Park, Keefe Ave. & Concordia in vicinity of tennis courts	126	66	76	M. Streufert (1975)	M. Streufert (1975)
Milwaukee	Milwaukee- Kern Park, Keefe Ave. & Concordia in vicinity of tennis courts	114	72	71	S E Roesch (1983)	S E Roesch (1983)

Populus alba White Poplar

County	Location	CBH	H	S	Last Measured	Nominator
Brown	Green Bay- 660 Florist Dr.	173	60	68	Jessica Schmidt (2004)	Paul Hartman (1978) 158" CBH
Waukesha	Oconomowoc- 60 Forrest Dr backyard	157	90	80	Wm Dahlquist & James Hovland (1972)	Wm Dahlquist & James Hovland (1972)

Populus balsamifera Balsam Poplar

County	Location	CBH	H	S	Last Measured	Nominator
Langlade	Town of Polar- T30N R12E Sec. 14 NE SW, 120 ft NW of bendin Red River in SW corner of 40	95	100	38	Gene Francisco (1976)	Gene Francisco (1976)
Lincoln	Merrill- 1100 Marc Dr. Merrill Stange's Park corner of W 3rd & Parkway	80	71	32	Jeff Barkley (1983)	Jeff Barkley (1983)

County	Location	CBH	H	S	Last Measured	Nominator

Populus deltoides

Eastern Poplar (Cottonwood)

County	Location	CBH	H	S	Last Measured	Nominator
Dodge	Columbus- 3 mi. N on Hwy 73, W 1/4 mi. on Kirchberg Rd., in farm field to N	360	97	101	R. Bruce Allison (2004)	R. Thomas & John Crombie (1976)
Green Lake	Town of Seneca- T17N R12E Sec. 26 NW NE, on N bank of Fox River approx. 2.5 mi from White River locks	354	110	99	T. Eddy & T. Jankowski (1987)	T. Eddy & T. Jankowski (1987)
Walworth	Sharon- W8071 Town Hall Rd., near airport, N42° 31.474' W088° 38.911'	346	120	98	David Farina (2002)	Mike Cerny (2000) 333" CBH
Iowa	Linden Township, T5N R2E Sec. 6, 1/4 mi E on Bucket Rd. S of fenceline	328	80	80	Clifford Siebert & Richard S (1988)	Clifford Siebert & Richard S (1988)

Populus grandidentata

Bigtooth Aspen

County	Location	CBH	H	S	Last Measured	Nominator
Vilas	Boulder Junction- 4125 CTH M, T43N R05E Sec. 26 b/w NW NE & NE EW, about 200-300' N of Papoose Lake Rd	137	87	56	K & B-W Hoffmann (2004)	Jeffery Olson (2002)
Iron	T41N R4E Sec. 36 NE1/4 NE 1/4, near 1st bridge on W River Trail	104	108	41	Al Murray (1993)	Al Murray (1993)
Monroe	Town of Wellington- T15N R1W Sec. 19 SE SW along E boundary 616 yds N of SE corner of SE SW	101	105	36	Jack Halbrehder (1986)	Jack Halbrehder (1986)

Populus deltoides (Eastern Poplar)
Columbus
Dodge County

Populus deltoides (Eastern Poplar)
Sharon
Walworth County

Populus grandidentata (Bigtooth Aspen)
Boulder Junction
Vilas County

County	Location	CBH	H	S	Last Measured	Nominator

Populus nigra   Lombardy Poplar

County	Location	CBH	H	S	Last Measured	Nominator
Door	Ephraim Township-1/4 Mile E. Hwy 42	203	60		Dan Crooks (1990)	Dan Crooks (1990)
Bayfield	Washburn- Hwy 13, Dairy Queen	168	78	35	(1989)	(1989)

Populus tremuloides   Quaking Aspen

County	Location	CBH	H	S	Last Measured	Nominator
Ashland	Town of Morse-T44N R3W Sec. 11 SW SENW next to woods trail	123	80	45	Steve Courtney (1994)	Allan N. Cramer (1994)
Clark	Town of Green Grove- T28N R1W Sec. 21 SW NE	103	97	41	Stan Schultz & Sons (1987)	Stan Schultz & Sons (1987)
Chippewa	Town of Cleveland-T31N R7W Sec. 4 on Moonridge Trl	95	104	43	(1983)	Richard Lindberg (1961)
Price	Flambeau- T39N R2W Sec. 14 NE NW	92	83	36	David Klug (1983)	David Klug (1983)

Prunus armeniaca   Apricot

County	Location	CBH	H	S	Last Measured	Nominator
Brown	Green Bay-724 St. George St	38	24	41	H. Plansky (2004)	P. Hartman (1994) 34" CBH

County	Location	CBH	H	S	Last Measured	Nominator

Prunus serotina — Black Cherry

County	Location	CBH	H	S	Last Measured	Nominator
Waushara	Town of Richford-T18N R9E Sec. 3 NE NE	125	75	35	Michael Bednarek (1994)	Michael Bednarek (1986)
Shawano	Aniwa Township-T29N R11E Sec. 1 SW NW, 70 yds E of Cedar Rd. down a two-track woods road	115	89	56	Larry Forden (1991)	John Stengl (1991)
Columbia	Baraboo- E10326 CTH W, T11N R05E Sec. 08 SWNE NE, 300 yds S of Bender farm along side sheep pasture	109	103	43	Rick Livingston (2002)	Rick Livingston (2002)

Prunus virginiana — Common Chokecherry

County	Location	CBH	H	S	Last Measured	Nominator
Fond du Lac	Waupun- Hwy AW W14285, 6 mi W of Waupun, N side on road on top of high knoll in field	65	46	33	Rodger Schley (2000)	Rodger Schley (2000)
Sheboygan	Sheboygan-1420 N 12th St	46			D. P. Grant (1988)	D. P. Grant (1988)

Pseudotsuga menziesii — Douglasfir

County	Location	CBH	H	S	Last Measured	Nominator
Dane	Madison-1007 Hillside Ave.	84	50	25	R. Bruce Allison (2004)	Terry Kelly (2004)
Brown	Allouez- 1542 Webster Ave, NE of Woodlawn Cemetery	80	90	35	B. Lange (2004)	T. Lang (1994) 72" CBH

Pseudotsuga menziesii (Douglasfir)
Madison
Dane County

Quercus alba (White Oak)
Fitchburg
Dane County

County	Location	CBH	H	S	Last Measured	Nominator

Pyrus communis
Common Pear

County	Location	CBH	H	S	Last Measured	Nominator
Waukesha	Pewaukee Village- N28 W25206 Bluemond Rd	95	35	30	C Harrington (1980)	C Harrington (1980)
Manitowoc	Franklin Township- T20N R22E Sec. 15 SW1/4 SW1/4	90	39	37	Wayne Norris (1990)	Wayne Norris (1990)

Quercus alba
White Oak

County	Location	CBH	H	S	Last Measured	Nominator
Dane	Fitchburg- 5774 Adams Rd, .4 mi W of Hwy D, 150 yds. N of road	202	76	69	R. Bruce Allison (2004)	W. E. Scott (1979)
Waukesha	Oconomowoc- N79 W33693 Peterson Rd, behind home in the woods	198	96	69	Stan Binnie (1996)	Stan Binnie (1996)
Iowa	On CTY K approx 4 mi S of U.S. 14, in front yard 50' from road	190	61	97	Rovert Van Pelt (1983)	Rovert Van Pelt (1983)
Sheboygan	Town of Lyndon- T14N R21EW Sec.34 1/2 NE SW	188	98	84	Lawrence E. Baer (1995)	Al Zuengler (1983)
Walworth	Burlington- 35303 Miller Road, half mi down driveway	181	80	70	Darryl Craig (2002)	Darryl Craig (2002)

County	Location	CBH	H	S	Last Measured	Nominator

Quercus bicolor   Swamp White Oak

County	Location	CBH	H	S	Last Measured	Nominator
Juneau	Town of Finely- Sec. 10 NE & located E of Yellow River by 200-300'	182	81	89	Daniel Mielke & Adam Mielke (1994)	Daniel Mielke & Adam Mielke (1994)
Waukesha	Oconomowoc- 120 Lac La Belle Ct	168	50	100	James F Zahradka (1979)	James F Zahradka (1979)
Rusk	Town of Big Bend- T33N R8W Sec. 12 Govt Lot 1	163	80	50	Don Rhone (1989)	Don Rhone (1989)
Grant	Bagley- T5N R6W Sec. 18 SW 1/4 SW 1/4	156	90	69	Dan Oles & Loren Danson (1995)	Dan Oles & Loren Danson (1995)

Quercus coccinea   Scarlet Oak

County	Location	CBH	H	S	Last Measured	Nominator
Dane	Madison- UW- Shorewood, entrance to university homes, next to walk, UW ID# 32-393	102	60	50	R. Bruce Allison (2004)	B. Thomas (1975)

Quercus ellipsoidalis   Northern Pin Oak

County	Location	CBH	H	S	Last Measured	Nominator
Shawano	Shawano- W6569 Hwy 47/55, T27N R15E Sec 13, in driveway of farm house .4 mi N of CTH A on Hwy 47/55 on W side	174	78	55	S E Crowley (2000)	S E Crowley (2000)
Portage	Stevents Point- Freemont Street, at SE corner of UW Stevens Point Center	164	92	80	Robert Freckmann (2004)	Tim Yanacheck (2003)
Shawano	Krakow- W155 Angelica	163	78	87	Michael Maederer (2000)	Michael Maederer (2000)

Quercus coccinea (Scarlet Oak)
Madison
Dane County

Quercus ellipsoidalis (Northern Pin Oak)
Stevens Point
Portage County

Quercus macrocarpa (Bur Oak)
Dousman
Waukesha County

County	Location	CBH	H	S	Last Measured	Nominator

Quercus imbricaria   Shingle Oak

County	Location	CBH	H	S	Last Measured	Nominator
Rock	Beloit- 2348 Bultin Drive, back yard	100	77	68	Mary Ann Kroehn Buenzow & Chris Ranum (2004)	Fred Veihman (1989) 68" CBH
Walworth	Whiewater- UW Campus b/w Campus Center & Red School House	64	52	50	Steven Bertagnolli (2004)	Maurice Kalb (1981) 61" CBH

Quercus lyrata   Overcup Oak

County	Location	CBH	H	S	Last Measured	Nominator
Milwaukee	Milwaukee- 2405 W Forest Home Ave, Forest Home Cemetery near Edward Schmidt Monument	52			S. E. Roesch (1983)	S. E. Roesch (1983)

Quercus macrocarpa   Bur Oak

County	Location	CBH	H	S	Last Measured	Nominator
Waukesha	Dousman- W394 S4086 Hwy Z, Stone Fences Farm N42° 58.304' W88° 31.789'	249	55	80	R. Bruce Allison (2004)	Richard Northey (1985)
Kenosha	Town of Paris- farm yard tree next to machine shed	201	83	65	Mike Schneider (1991)	Mike Schneider (1991)
Lafayette	Blanchardville- 504 Mound St, T02N R05E Sec. 20 SE SE SW	193	89	85	Matt Singer (2001)	Den 4, Pack 125 Darlington Cub (2002)
Racine	Raymond Township- cemetery SE of inter- sect of CTH G & U	190	55	91	Michael Schneider (1988)	Michael Schneider (1988)

County	Location	CBH	H	S	Last Measured	Nominator

Quercus muehlenbergii   Chinkapin Oak

County	Location	CBH	H	S	Last Measured	Nominator
Waukesha	Menomonee Falls- North Hills Country Club, N73W13430 Appleton Ave., near #4 green	135	80	59	Mike Faust & B-W Hoffmann (2004)	P. Buckley (1983), Greens Crew of Country Club (1995) meas. 125
Milwaukee	Milwaukee- 4500 W Cluster Ave, McGovern Park	96	76	67	R Rideout (1983)	R Rideout (1983)
Grant	Town of Waterloo- T3N R4W Sec. 15 SE SE	80	48	67	Craig Hollingsworth (1989)	Craig Hollingsworth (1989)

Quercus palustris   Pin Oak

County	Location	CBH	H	S	Last Measured	Nominator
Dane	Madison- Camp Randall UW Campus, UW ID # 87-7942	138	70	75	Daniel Einstein (2004)	E R Hasselkus (1965)
Dane	Madison- Camp Randall UW Campus, UW ID# 87-7943	127	60	50	Daniel Einstein (2004)	T. Klitzkie (1975) 92" CBH
Dane	Madison- Camp Randall UW Campus, UW ID # 87-7941	118	50	45	Daniel Einstein (2004)	T. Klitzkie (1975) 87" CBH

Quercus prinus   Chestnut Oak

County	Location	CBH	H	S	Last Measured	Nominator
Dane	Madison- UW Houses # 35 near play area, UW ID# 31-394	67	40	40	Daniel Einstein (2004)	B. Thomas (1975) (1979) 55" CBH
Columbia	Town of Dekorra- T11N R9E Sec. 25, located W of Conservation Exhibit & S of fence	61	47	40	Derrick Duane (2004)	Kenneth W. Wood (1985) 46" CBH

Quercus muehlenbergii (Chinkapin Oak)
Menomonee Falls
Waukesha County

Quercus palustris (Pin Oak)
Madison
Dane County

Quercus prinus (Chestnut Oak)
Madison
Dane County

County	Location	CBH	H	S	Last Measured	Nominator

Quercus robur   English Oak

County	Location	CBH	H	S	Last Measured	Nominator
Milwaukee	Milwaukee- Forest Home Cemetery, 2405 Forest Home Ave.	92	73	41	Dale R Konieczka (2004)	Dale R Konieczka (1980)
Milwaukee	Milwaukee- Forest Home Cemetery, 2405 Forest Home Ave.	87	68	44	Dale R Konieczka (2004)	Dale R Konieczka (1980)

Quercus rubra   Red Oak

County	Location	CBH	H	S	Last Measured	Nominator
Dane	Madison- Mendota Mental Health Insititue near Eagle Effigy Mound N43° 07.804' W089° 23.844'	226	85	65	R. Bruce Allison (2004)	Walter E. Scott (1976)
Dane	Goose Lake SWA- Far eastern edge of Dane County, S side of lake, less then 100 yds E of Missouri Rd. 0.2 to 0.3 mi N of CTH BB	204	79	88	Bill Klein (2003)	Bill Klein (2003)
Sheboygan	Rhine- T16N R21E Sec. 17 NE NE	203	89	67	Roland Kuhn (1995)	Eleanor Kuhn (1985)
Wood	Aburndale Township- T25 R4E Sec. 13 SE SW, N side of Yellowstone Rd	195	84	65	Steve Grant (2002)	Steve Grant (2002)
Iowa	Avoca- 6084 Hwy 130, 300 yds S-SE of farmhouse W of walking trail	182	105	90	Tom Hill (2002)	DR Downs (2000)

Quercus robur (English Oak)
Milwaukee
Milwaukee County

Quercus rubra (Red Oak)
Madison
Dane County

Quercus velutina (Black Oak)
Albany
Green County

County	Location	CBH	H	S	Last Measured	Nominator

Quercus velutina   Black Oak

County	Location	CBH	H	S	Last Measured	Nominator
Green	Town of Albany-410 E State St., behind St. Patrick's Church N42° 42.251' W89° 25.950'	210	70	81	R. Bruce Allison (2004)	Boyd Atkinson
Waukesha	Town of Brookfield-19185 Timberline Dr. backyard	191	85	78	Mike Guth (2002)	Mike Guth (2002)
Green Lake	Town of Marquette-T14N R11E Sec. 34 NE	183	61	29	Tim Jankowski (1983)	Tim Jankowski (1983)

Robinia pseudoacacia   Black Locust

County	Location	CBH	H	S	Last Measured	Nominator
Milwaukee	Brown Deer-4920 Goodhope Rd.	188	60	40	Brian Buntroch (1995)	R. Thurow (1983)
Waushara	Wautoma- T18N R10E Sec. 3 SE NW, W side of Hwy 22	149	90	45	Michael Bednarek (1989)	Michael Bednarek (1986)

Salix alba   White Willow

County	Location	CBH	H	S	Last Measured	Nominator
Juneau	Lemonweir Township- T15N R4E Sec. 24 SW SE	290	63	81	M K McClurg (1985)	M K McClurg (1985)
Walworth	Lake Geneva- 60 Snake Rd along lake path N42° 34.941' W088° 27.046'	247	65	80	R. Bruce Allison (2004)	(1981) 230" CBH

Salix amygdaloides   Peachleaf Willow

County	Location	CBH	H	S	Last Measured	Nominator
Milwaukee	West Allis-Greenfield Park W side of 116th St	369	80	77	Eugene Zanow (1988)	Eugene Zanow (1988)

Salix alba (White Willow)
Lake Geneva
Walworth County

County	Location	CBH	H	S	Last Measured	Nominator

Salix babylonica   Weeping Willow

County	Location	CBH	H	S	Last Measured	Nominator
Pierce	Salem Township- Sec. 26, Rt. 2 Maiden Rock in yard	267	65	72	Joe & Lynette Traynor (1990)	Joe & Lynette Traynor (1990)
Ozaukee	Knollwood- 3439 Knollwood	205	55	70	Arthur R. Schuelke (1990)	Arthur R. Schuelke (1990)
Taylor	Medford- 602 Allman backyard	174	58	55	Michael J Riegert (1989)	Michael J Riegert (1989)
Outagamie	Town of Greenville- Sec. 2 at W6527 City Trunk JJ, S of garage	170	70	65	Bob Schroeder (2004)	Bob Schroeder (2004)

Salix discolor   Pussy Willow

County	Location	CBH	H	S	Last Measured	Nominator
Taylor	Stetsonville- 125 E Mink Ave, NW corner of lot, 100 yds E of Hwy 13	39	26	24	Michael J Riegert (1988)	Michael J Riegert (1988)

Salix fragilis   Crack Willow

County	Location	CBH	H	S	Last Measured	Nominator
Taylor	Medford- 621 Allman backyard	181	61	73	Michael J Riegert (1989)	Michael J Riegert (1989)

Salix matsudana   Corkscrew Willow

County	Location	CBH	H	S	Last Measured	Nominator
Brown	Green Bay- 1291 Oregon	162	60	54	M. Freberg & E. Muecke (2004)	M. Freberg (1994) 116" CBH
Milwaukee	Milwaukee- 1228 East Russell	106	51	46	(1983)	(1983)

Salix nigra   Black Willow

County	Location	CBH	H	S	Last Measured	Nominator
Fond du Lac	Town of Taycheeda- T16N R18E Sec. 17 SE NE	264	76	94	Ted Pyrek & Frank Hanson (1984)	Ted Pyrek & Frank Hanson (1984)

County	Location	CBH	H	S	Last Measured	Nominator

Salix x sepulcralis — Golden Weeping Willow

County	Location	CBH	H	S	Last Measured	Nominator
Brown	Green Bay- 1133 Division	210	78	79	M. Freberg & E. Muecke (2004)	Sue Baumgartel (1994) 186" CBH

Sassafras albidum — Common Sassafras

County	Location	CBH	H	S	Last Measured	Nominator
Dane	Madison- 1 Thorstrand Rd b/w house & lake	49	35	30	R. Bruce Allison (2004)	W. E. Scott (1975) 28" CBH
Kenosha	Kenosha- 907 71st St back yard	28			B. Sanfelippo (1978)	B. Sanfelippo (1978)

Sorbus americana — American Mountainash

County	Location	CBH	H	S	Last Measured	Nominator
Sauk	At Lakeway Resort, corner of Hwy 123 & Gall Rd, 1 mi S of Baraboo	72	48	41	Chris Goodwin & John Sauer (2004)	James Kremsreiter (1986) 53" CBH
Trempealeau	Town of Preston- T21N R7W Sec. 30 SW SW	58	60	33	Gary Zielske (1986)	Gary Zielske (1986)
Portage	Stevens Point- 15' W of N entrance to Allen Student Center	35	29	26	Mark Kinschi (1985)	Mark Kinschi (1985)

Sorbus aucuparia — European Mountainash

County	Location	CBH	H	S	Last Measured	Nominator
Wood	Marshfield- 514 W Arnold St., Mrs. Anna Grewohl Residence	73	33	28	Chad McGrath (1981)	Chad McGrath (1981)
Kewaunee	Kewaunee- 523 Vliet St	66			Jim Shefcher (1981)	Jim Shefcher (1981)
Kewaunee	Kewaunee- 522 Wisconsin Ave	55			Mr. & Mrs. Thomas Vogel (1981)	Mr. & Mrs. Thomas Vogel (1981)

Sassafras albidum (Common Sassafras)
Madison
Dane County

County	Location	CBH	H	S	Last Measured	Nominator

Syringa pekinensis · Chinese Tree Lilac

County	Location	CBH	H	S	Last Measured	Nominator
Kenosha	Kenosha- 6541 7th Ave	67	40	41	(1988)	(1988)
Walworth	Whitewater- UW Whitewater in front of Old Main Hall	39	25	25	Steven Bertagnolli (2004)	E. Hasselkus (1975)
Brown	Green Bay- Jackson Park	37	27	25	T. Barber (2004)	P Hartman (1980) 31" CBH

Syringa reticulata · Japanese Tree Lilac

County	Location	CBH	H	S	Last Measured	Nominator
Shawano	Shawano- 225 E. Division St	70	35	27	(1973)	(1973)
Walworth	Whitewater- 133 S Cottage St., Richard Heidenreich Res.	69			Ruth Miner (1980)	Ruth Miner (1980)
Brown	Green Bay- 1215 Eliza	60	28	24	H. Plansky (2004)	P. Hartman (1978)
Outagamie	Appleton- 537 N Union St.- backyard	50	78	47	Bob Schroeder (2004)	David Z Rozmarynoski (1987)
Dane	Mt. Horeb- 106 S 5th St.	48	34	32	Jeff Gorman (2004)	Jeff Gorman (2004)

Syringa vulgaris · Common Lilac

County	Location	CBH	H	S	Last Measured	Nominator
Waukesha	Oconomowoc- Wisconsin Ave in front of a house	61	20	15	(1981)	(1981)
Brown	Green Bay- 330 S Jackson	47	23	20	T. Barber (2004)	P Hartman (1980)

Tamarix ramosissima · Tamarisk

County	Location	CBH	H	S	Last Measured	Nominator
Brown	DePere- 502 Suburban Dr.	12	16	18	D. Melichar (2004)	PJ Holschbach (1994) 9" CBH

County	Location	CBH	H	S	Last Measured	Nominator

Taxodium distichum — Common Baldcypress

County	Location	CBH	H	S	Last Measured	Nominator
Kenosha	Brighton Township- T2N R20E Sec. 10 NW1/4 SW1/4, Bong Recreation Area, 42° 38.70'N 88° 07.797'W, E side of access rd	135	50	51	R. Bruce Allison (2004)	Phil Sander (1962)
Dane	Madison- 4110 Mandan Crescent	129	65	35	R. Bruce Allison (2004)	M. A. Boettger (1973)
Walworth	Lake Geneva- W2765 S. Lakeshore Dr., next to Gatehouse, former- ly 840 Lakeshore Dr., N42° 24.464' W88° 26.196'	101	50	55	R. Bruce Allison (2004)	E. R. Hasselkus (1981) 75" CBH
Dodge	Beaver Dam- 205 S. Lincoln Ave, Lakeview Park by NW corner of tennis court	65	33	33	Ranger Services (2002)	Brian Pelot (2002)

Taxus cuspidata — Japanese Yew

County	Location	CBH	H	S	Last Measured	Nominator
Brown	Green Bay- 840 S. Madison	63	38	24	M. Freberg & E. Muecke (2004)	David Crawford & M. Freberg (1994) 55" CBH
Dane	Maple Bluff- 221 Lakewood Blvd N43 06.530 W89 22.036	62	25	35	R. Bruce Allison (2004)	M. Schneider (1980) 45" CBH
Kenosha	Kenosha- 7402 Sheridan Rd	58	23	28	(1988)	(1988)
Kenosha	Kenosha- 3703 Sheridan Rd, Orzolek Res.	47	22	31	M Schneider (1988)	M Schneider (1988)

Taxodium distichum (Baldcypress)
Brighton Township
Kenosha County

Taxodium distichum (Baldcypress)
Madison
Dane County

Taxodium distichum (Baldcypress)
Lake Geneva
Walworth County

County	Location	CBH	H	S	Last Measured	Nominator

Thuja occidentalis — American Arborvitae

County	Location	CBH	H	S	Last Measured	Nominator
Douglas	Town of Summit- T47N Sec. 29 NW NW, 1/4 mi E of state line near jct of N & S forks of Nemadji River	161	89	41	Dr. Scott Nielson (1986)	Dr. Scott Nielson (1986)
Manitowoc	Town of Gibson- T21N R23E Sec. SE SE	160	87	28.5	R. Pleuss (1987)	Raymond Vesely (1985)

Tilia americana — Amercian Linden (Basswood)

County	Location	CBH	H	S	Last Measured	Nominator
Outagamie	Town of Greenville- Sec. 2, 1/2 mi. S of JJ, 100' W of creek	276	85	63	Bob Schroeder (2004)	Bob Schroeder (2004)
Forest	Town of Hiles- T40N R12E Sec. 28 SW NE, approx 200' from Crossover Rd. & .25 mi S of Cody Resort	182	112	49	William H. Dixon (1984)	William H. Dixon (1984)

Tilia cordata — Littleleaf European Linden

County	Location	CBH	H	S	Last Measured	Nominator
Dane	Middleton- 7426 Elmwood Ave, street tree	137	60	45	R. Bruce Allison (2004)	(1983) 109" CBH
Brown	Green Bay- 1608 Beauchamp	129	47	49	M. Freberg & E. Muecke (2004)	M. Freberg (1993) 94" CBH
Dane	Madison- 28 Lathrop St backyard	119	60	45	R. Bruce Allison (2004)	E. Hasselkus (1971)
Milwaukee	Milwaukee- Washinton Park Picnic Area #5 near tennis courts	104	60	44	R. Rideout (1987)	R Hodkiewicz (1984)

Tilia cordata (Littleleaf Linden)
Middleton
Dane County

County	Location	CBH	H	S	Last Measured	Nominator

Tsuga canadensis   Canadian/Eastern Hemlock

County	Location	CBH	H	S	Last Measured	Nominator
Bayfield	Town of Clover- T50N R7W Sec. 24 Center, approx 1/2 mi SE of river	153	96	49	Scott Nielson (1986)	Scott Nielson (1986)
Marathon	Town of Easton- T29N R9E Sec. 18 SW	129.6	94	43	Charles Newby (1988)	Charles Newby (1988)
Iron	T44N R1E Sec. 13 SE SE	121	95	40	Rudy Kangas (1990)	Rudy Kangas (1990)
LaCrosse	West Salem- Hamilton Cemetery adjacent to Hwy 16 & M	95	78	52	Chris Winther (1986)	Chris Winther (1986)

Ulmus americana   American Elm

County	Location	CBH	H	S	Last Measured	Nominator
Outagamie	Town of Ellington- Sec. 24 on Rock Rd E of Greenwood Town of Greenville, Sec 2 about 1/2 mile S of JJ, 100' W of creek	228	105	70	Bob Schroeder (2004)	Bob Schroeder (2004)
LaCrosse	Town of Greenfield- T15N R6W Sec. 5	216	75	100	R. Machotka (1986)	J. Davis
Jefferson	Ixonia- T8N R15E Sec. 5, St. Paul's Cemetary on CTH SC, 1.5 mi N of STH 16, 1 mi N of CTH R	215	79	108	David Schumann (2001)	David Schumann (2001)
Waupaca	Little Wolf Township- T23N R13 E Sec. 29 SW NE, Spring Creek Rd. S of CTH B, across from Spring Brook School	214	70	98	Shannon Lettau (1991)	Shannon Lettau (1991)
Green	Mount Pleasant Township- T3N R8E Sec. 26 NE1/4, W side of Rechsteiner Rd. b/w Purinton Rd & Hwy 39	197	70	92	Mark K. Leach & Jon Kollitz (1988)	Mark K. Leach & Jon Kollitz (1988)

County	Location	CBH	H	S	Last Measured	Nominator

Ulmus glabra   Scots Elm

County	Location	CBH	H	S	Last Measured	Nominator
Milwaukee	Milwaukee- 3312 N Lake Dr	79	65	48	E Hasselkus & R Rideout (1986)	E Hasselkus & R Rideout (1986)

Ulmus procera   Engish Elm

County	Location	CBH	H	S	Last Measured	Nominator
Walworth	Lake Geneva- 930 Bayview Dr, tennis courts	130			E. Hasselkus (1981)	E. Hasselkus (1981)
Milwaukee	Milwaukee- 2405 W Forest Home Ave, Forest Home Cemetery, near Meckelburg Stone	110			S. E. Roesch (1983)	S. E. Roesch (1983)

Ulmus pumila   Siberian Elm

County	Location	CBH	H	S	Last Measured	Nominator
Milwaukee	Milwaukee- 3734 W Mitchell St	200	89	85	R. Bruce Allison (2004)	J Kringer (1983)
Fond Du Lac	Osceola Towhship- T14N R19E Sec. 1 NE SE, along trail	124	43	74	Royal Carey (1987)	Royal Carey (1987)

Ulmus rubra   Slippery Elm

County	Location	CBH	H	S	Last Measured	Nominator
Grant	Platteville- edge of woods in stream flood plain, Bell's woods	124			W. E. Scott (1980)	W. E. Scott (1980)

Ulmus pumila (Siberian Elm)
Milwaukee
Milwaukee County

County	Location	CBH	H	S	Last Measured	Nominator

Ulmus thomasi Rock Elm

County	Location	CBH	H	S	Last Measured	Nominator
Jefferson	Ft. Atkinson- W4427 HWY 106, T15N R06E Sec. 32 NW E NW SW, E side of driveway	151	65	70	Greg Kirchmayer (2001)	Ronald Hall (2001)

Ulmus x glabra Camperdown Elm

County	Location	CBH	H	S	Last Measured	Nominator
Racine	Burlington- NE corner of Crossway & Ketterhaugen, 94 to Hwy 142 & past 45 & 75 on 142, go through Bong State Park	100	37	41	Thomas F. Cummings (1995)	David & Joyce Nilles (1995)
Kenosha	Wilmot- 11535 304 Ave, Home of Robert Ehlert	99	24	31	(1985)	Walter E. Scott (1963)
Kenosha	Kenosha- 8301 104th Ave, 40' S of house	80	20	29	Tom Cummings (1986)	Tom Cummings (1986)

Viburnum lentago Nannyberry Viburnum

County	Location	CBH	H	S	Last Measured	Nominator
Dane	Black Earth- Box 4621 Old Indian Trail next to driveway	30			R. Sonnenberg (1981)	R. Sonnenberg (1981)
Winnebago	Oshkosh- 327 E. Irving Ave next to driveway	22			G. Herold (1978)	G. Herold (1978)

COMMON NAME INDEX